The HCG Diet

Everything You Need to Know about The HCG Diet and More…

G. McGwire

The HCG Diet

Copyright © 2011 by G. McGwire

Table of Contents

What Is the hCG Diet?

The Facts Behind the Latest Diet Craze

It's been on the covers of magazines, promoted online, and you've heard that some of your favorite celebrities have tried it – but what is the hCG diet? Where do you go to find out more? In this book, you'll learn all you need to know about the hCG diet.

We'll start with the basics and continue with how you can maintain the success you've experienced. You'll go from a novice to an expert as you decide if this diet is right for you and what you need to do to get started.

Let's start at the very beginning and look at what hCG stands for, what it is, and what this diet plan is really all about.

What Is hCG?

hCG is actually a hormone called Human Chorionic Gonadotropin. With a name like that, it's no wonder it's been shortened to hCG. This is a hormone that actually gets produced in high levels during pregnancy.

This hormone is produced by the placenta in a female's body. Once it starts being produced, it takes over the metabolism of the pregnant woman. It tends to be found in high levels in early pregnancy and lower levels as the pregnancy develops.

You may be thinking, "But women gain a whole lot of weight during pregnancy. How can this work to lose weight?" In actuality, hCG doesn't cause anyone to gain weight. Instead, it helps you body to use stored fat as energy.

Burning the Right Type of Fat

When you think about losing weight, you usually think that getting fat off of your body is the most important thing. But there's more than one type of fat – and your body needs to keep some types of fat in order to be healthy.

Some fat is in your body as a protection against shock and injury. It helps to protect and cushion your joints and vital organs. You want to leave this fat alone – it has an important job to do. In fact, if you have too little fat it can be detrimental to your health.

Then there's fat that helps you keep up with your body's dietary needs. When you eat more calories than you need, your body will store the extra as fat. This is really important if you have a day (or a few days) when you can't get enough calories.

You need to have some fat storage for energy reserve. The problem is that many people have so much fat stored that it becomes unhealthy. When you have too much fat, the effects can be detrimental.

Excess fat can become wrapped around your internal organs. This is like having too much oil buildup in your car. It slows down the system making it inefficient. It can also alter your metabolism and cause you to slow down the rate at which you burn calories.

Having too many extra pounds around your middle can increase your risk of heart disease and stroke. All that fat has to have a blood supply and it's taxing to your heart to circulate the blood through extra tissues.

Fat can also become deposited on your arteries and put your at higher risk for a heart attack. A link has also been found between obesity and cancer. Increased unhealthy fat can actually cause your cells to grow in abnormal ways.

When it comes to dieting, you want to burn this type of fat – the fat found in your internal organs and especially the fat that settles in your midsection. The hCG diet targets body fat stores and helps you to burn them for your daily energy.

In a normal diet where you only restrict calories, your body will start to burn the structural fat that you need to stay healthy instead of targeting the unhealthy fat. In fact, your body will only burn the unhealthy fat after you've put your body into a starvation mode or during pregnancy.

That's where hCG comes in. With this program, your body will use the abnormal fat as its first source for calorie burning. Instead of attacking the fat that keeps you healthy, you'll be targeting the fat that's harming you.

hCG Releases Fat Stores

Adding hCG to your diet program will help to boost your fat burning powers. The hormone actually causes your body to use stored fat deposits as energy. What that means for you is that you can shed pounds quickly.

You'll be eating a low calorie diet (we'll get into the specifics in Chapter 2), but instead of thinking it's starving, your body will burn fat. And you won't even feel hungry because you'll have plenty of energy from the stored fat.

Where Does hCG Come From?

Now that you know what hCG is, you're probably wondering how you can add it to your body. If people produce this during pregnancy, is it possible to take it when you're not pregnant? While hCG is a hormone produced in pregnancy, there are a few options for adding it when you're not pregnant.

It can be taken in the form of injections or in sublingual (under the tongue) drops. These supplements typically come from synthetic sources – not from humans or animals. When you decide to get started, you'll need to decide between drops and injections.

Men and the hCG Diet

Because hCG is a hormone associated with pregnancy, many men are unsure of its safety. However, men are just as responsive to the hCG diet as women. Men have just as much success on the hCG diet as women. In fact, they generally have fewer side effects from the hormone.

Don't let the fact that this hormone is associated with pregnancy scare you. While this hormone is found in higher levels in pregnant women, it's found in both men and women at low levels. Adding hCG to your regimen is not going to affect your gender.

How Much Weight You Can Expect to Lose

When it comes to weight loss, the hCG diet stands out among other competitors. With a typical diet where you restrict calories and add exercise to your lifestyle, you can expect to lose about 5-7 pounds the first week and then 1-2 pounds per week after that.

The hCG diet will introduce you to a whole new rate of weight loss. Typically, a person following the plan will lose an average of 1-2 pounds per day. That means in 30 days you could lose 30 pounds.

And what's even better, that weight loss will come from actual fat loss. Instead of losing only water weight and muscle mass as most diets do, you'll actually lose only from your fat stores. You'll begin to have the body you've always dreamed of.

You may have heard that this type of rapid weight loss is not safe. However, with the hCG diet, those fears are unfounded. Because you are only burning excess fat, you won't fall to the normal weight loss pitfalls including muscle loss, bone loss, and vitamin loss.

Instead of worrying about malnutrition, you'll just need to worry about how you're going to plan for your new wardrobe! With the hCG diet, you'll lose fat and maintain your good health. This is a plan for those who don't want to wait for slow weight loss methods to work.

Drops vs. Injections

There are two different delivery methods – injections or sublingual drops. There are pros and cons to each. One of the first big decisions you'll need to make is which of these you'd like to try.

Injections of hCG hormone are given only by prescription from a physician. You'll have to go to the doctor to get these prescribed and you'll continue to be monitored by that doctor. Your health insurance isn't going to cover this product, so you'll have to pocket the expense.

For some people, it gives peace of mind to go to a doctor and to know that they will be monitored carefully. However, some people object to the idea of taking a shot every day – and even worse giving themselves a shot.

Injecting yourself with a shot takes a bit of practice and, obviously, it hurts. In addition you'll have the added expense of syringes, sharps containers, alcohol swabs, and bandages. For some people this is a deterrent while others don't have a problem with it.

Sublingual drops, on the other hand, are available without a prescription from homeopathic suppliers. You can purchase them online or in some natural products stores. They are drops given daily under the tongue.

While you can certainly have your physician monitor you while on this diet – and in many cases that's a good idea – you aren't required to have a physician write a prescription. That makes this diet more accessible for people who don't want to bother with going to the doctor.

Taking drops under the tongue is a very efficient delivery method for supplements – it's a pathway directly to your bloodstream. In fact, there are many medications that are given sublingually instead of through injections.

This method is less invasive, less painful, and less intimidating. You'll have to make up your own mind about what's right for you. And before you start any diet program, it's a good idea to speak with your healthcare provider and get the okay to begin.

Ready to Find Out More?

Now that you know the basics of hCG and what it does in the body, you're probably anxious to find out more about the diet plan. In this book, you'll learn all about the diet plan and what you need to do to follow it. You'll also learn what side effects you may want to watch out for.

In addition, you'll learn how exercise can affect your progress on the hCG diet and how to set an appropriate goal weight for your diet program. You'll even get information on how to beat plateaus that threaten to slow down your success.

The hCG diet is a program for people who want to lose as little as 10 pounds and even for those who have more than 100 pounds to lose. As you read, you'll learn how to properly lose the fat for good.

Anyone can lose a few pounds for a special event or to fit into that dress they've always wanted to wear. But are you ready to lose weight and keep it off permanently? If you're looking for a plan that will allow you to totally reset your metabolism and burn fat, the hCG diet can deliver.

In the next few chapters you'll learn all you need to do to finally have diet success and the body you've always wanted. Not to mention you'll have increased vitality and good health. It's time to learn more about the hCG diet plan.

The hCG Diet Step by Step

Breaking Down the Diet Plan

If you're interested in losing weight fast and keeping it off permanently, the hCG diet is a great option. We're going to discuss how you actually follow the diet plan. You'll learn everything from what to eat, to how often to take hCG, to how many cycles you need to follow.

The hCG diet actually is followed in cycles. You'll stay on the plan for 21-42 days depending on the desired weight loss. Then you'll take a break and go into a maintenance phase. Later you can follow another cycle to lose more weight.

Phase One – Ramping Up the Burn

The first two days on the hCG diet are essential for ramping up your fat burning capabilities. You'll want to follow these two days to the letter because they are going to set the stage for the amazing weight loss you're going to have.

During this time you'll start you hCG supplements. Whether you use injections or drops, you'll follow the same protocol. At this time you'll also begin following a high fat, high calorie diet. This may seem counterintuitive, but this is only for a couple of days while you begin the fat burning process.

With this program you're actually resetting your body's metabolism, so you'll notice that it's unlike anything you've tried before. You may have to fight your instincts at first, but realize that there's a method to this program that will work for you.

When you're considering what to eat, think about all those foods you've been forbidden to eat in the past – pork rinds, beef, chocolate cake, etc. Load up on those calories and fat for a couple of days and you'll set the stage for the hCG diet to work its magic.

You'll also need to start increasing your water intake. This is to allow the hCG hormone to move freely within your body and get the job of fat burning done. You'll need to drink at least 100 ounces of water each day beginning from the first day.

It may be hard to drink that much water if you're not used to it yet, so be prepared for the inevitable trips to the bathroom. Just plan to keep a glass or bottle of water with you at all times when you're following this plan.

When you've finished with the first two days, it's time to move on to the next phase. During this phase, you'll eat much fewer calories and your diet will be quite restrictive. However, you'll only have to follow this plan for 20-40 days.

Phase Two – Calorie Restriction and Power Weight Loss

After a couple of days eating a high fat diet, drinking large amounts of water, and taking hCG supplements your body is ready to begin really burning fat. Now it's time to decrease your caloric intake.

In fact, you'll only be eating 500 calories per day during this part of the diet. How can you live on such a low number of calories? Traditional diets have warned you that eating too few calories can trigger your body to decrease its metabolism.

But with the hCG diet, the hormone actually takes over your metabolism. You'll begin to burn only fat stores as you decrease your calories to 500 each day. You may feel hungry for the first couple of days, but soon that feeling will subside.

You won't feel as hungry because your body is going to use the stored fat as fuel. You won't crave food to give your body energy. But you'll need to continue eating food in order to keep your digestive system moving properly.

You'll begin experiencing major weight loss during this phase. You may burn as much as 2 pounds per day. Some people burn as much as 3 pounds per day. While it seems like a drastic change, you'll feel great when you see the results.

So what can you eat when you have only 500 calories to work with? You'll want to focus on eating lean meats and other proteins as well as eating vegetables and fruits. For example, each day you'll want to have:

- 1 serving of fruit, 2 times each day
- 1 serving of vegetables, 2 times each day
- 3 oz cooked lean meat, 2 time each day
- 2 melba toasts
- The juice of one lemon each day
- You can also have plenty of fresh herbs and seasonings

You'll find that while this seems restrictive, you can actually enjoy many of your favorite flavors. You'll want to avoid any butter, oil, or salad dressings as these add extra calories and fat.

During phase two, you want to have very little added fat – instead you want your body to used its stored fat. You'll also want to remember to keep drinking those fluids. It's a good idea to have ½ gallon to a gallon of water every day.

In addition to water, you can also drink coffee and tea. If you must have sweetener, use a sweetener that has no calories such as stevia. You'll also need to forego the cream in your coffee. If you absolutely must have cream, you may have up to 1 tablespoon of milk.

In addition to staying away from fats and oils, you'll also need to steer clear of breads and other starchy carbohydrates. Never fear, you'll be able to have them after you begin the maintenance phase – it's only a temporary separation.

Phase Three - Maintenance

After you've finished phase two, you're ready to begin the maintenance period. Remember you've only been eating 500 calories a day and it will take time for your body to get used to eating larger meals again.

During this time you'll slowly increase your caloric intake until you reach a more normal amount of calories. With most diets, this is when you begin to put weight right back on. But with the hCG diet you're body's metabolism will be reset.

Rather than putting on more weight, your body will continue to burn calories and you'll be able to eat a normal amount of food without gaining the weight you lost. When it comes to some foods, you'll need to take it very slowly.

For example, before adding sugars and starches back to your diet you'll need to wait two weeks. After two or three weeks you can begin slowly adding them back. Starches and sugars include foods like:

- Bread
- Cookies
- Candy
- Cereal
- Potatoes
- Rice

Remember to go slowly as these foods will be a shock to your cleansed system. In order to determine how much you can add without gaining weight, you'll want to weight yourself daily. If you find that you've gained more than two pounds, you'll need to cut back on these foods.

What If I Want to Lose More Weight?

A three week course of the hCG diet will help you to lose up to 15 pounds. This is great if you're trying to lose one or two dress sizes. A six week course will allow you to lose approximately 40 pounds. But what about people who have more to lose?

It's possible to lose more weight with the hCG diet, but you'll need to take a break from it. You'll need to wait at least six weeks before going back onto the plan. After six weeks, you can begin the hCG protocol again.

This will allow you to lose even more weight than you did the first round. For people who have a great deal of weight to lose, this is particularly good news. For most people, a goal weight can be obtained within a few months.

If you're ready to go back for a third or fourth course of the hCG diet, you'll need to wait even longer between courses. Between the second and third course you'll need to wait eight weeks and between a third and fourth course you'll need to allow 12 weeks.

Between the fourth and fifth course, you'll want to wait at least 20 weeks and if you want to finis a sixth course, you'll need to wait six months after the fifth to begin. Expect to follow this type of plan if you have substantial weight to lose.

Adapting for Vegetarians

There are some people who have special dietary needs that can make the hCG diet a challenge. For example, people who are restricted to a vegetarian diet may find that their weight loss is much slower.

You may want to get protein through milk or cheese, but these sources will not allow your body to burn fat as quickly. In fact, vegetarians usually only lose about half as much weight as their meat-eating counter-parts.

Instead, if you're a vegetarian you'll want to focus on sources of protein such as tofu, bean curd, and tempeh. You may also rely on legumes to help supplement your protein needs. Still, don't be surprised if your weight loss isn't quite as fast as it would be eating meat.

What If I Cheat?

Many people find dieting to be very difficult. While you won't feel hungry on the hCG diet after a few days, you may have some of your psychological demons to battle. If you find that you're feeling too restricted with the hCG diet, you may have the desire to cheat.

When you stray, even a little, from the plan you'll find that your weight loss is substantially slowed. Doing everything you can to be pre-pared ahead of time will reduce your chances of cheating.

Check out Chapter 10 for some ideas of foods that seem decadent, but are within the confines of the hCG plan. Remember, you only have to follow the very low calorie diet for a maximum of six weeks and then you can begin to return to normal.

Focusing on your weight loss goal and keeping the end in mind will make it easier for you to stick to the plan. You'll also want to celebrate the progress you're making. Sometimes seeing the numbers go down on that scale is enough to motivate you to hold on a few more days.

Will I Keep Off the Weight?

The biggest question most people have is, "Will I maintain my weight loss?" After all, many of us have lost weight only to find it creep right back up again. With the hCG diet, most people maintain their weight loss.

During the time you are following a very low calorie diet and taking the hCG hormone, your body is making big changes. You'll find that you aren't as hungry, that your tastes have changed, and that you're feeling more energetic than before.

This is a great time to institute lifestyle changes that will help you to keep off the weight for good. For many people this comes naturally as a result of following the diet plan for a course of several weeks.

You may also find that having lost weight you have more energy to exercise. When you are finished with a course of the hCG diet, you may want to add exercise to your program. In Chapter Five we'll discuss how to handle an exercise program.

Most people who follow this plan do keep the weight off. If you're one of the few who starts to see the scale go back up, you can always follow another course of the program and focus more on lifestyle changes as you enter the maintenance phase.

Making a Decision

As you can see, the hCG diet has a pretty strict regimen that must be followed. While some people thrive from the structure of the plan and the daily monitoring, other people find that this diet plan is too restrictive for their lifestyle.

Still other people may not want to follow the hCG plan because of health concerns and other restrictions by their physician. In the next chapter you'll learn about who should and should not consider following this plan.

For most people, the hCG diet is a very safe method for resetting the metabolism and taking weight off for good. It's a short protocol that has long lasting results. If you're ready to take the next step toward your weight loss goal, read on to find out if it's the right plan for you.

Weighing the Options

Is the hCG Diet Right for You?

If you're ready to lose weight and keep it off, the hCG diet plan can provide you with everything you need. And while most people can safely use this program, it isn't for everyone. In this chapter you'll learn about some special consideration with the hCG diet.

Once you know the facts, you can take all of that information and make an informed decision about what's best for you. Ultimately, you are the only one who can decide if the hCG plan is the best fit for what you need and what is safe for you.

Striving for Structure

The actual course of the hCG diet can be quite restrictive. It's essential that you follow the plan to the letter. If you are someone who is able to take direction well and follow specific instructions, the hCG diet will be a piece of cake for you.

But if weighing yourself daily and eating only very specific foods is unimaginable, this may not be the right course of action for you. You can always try a course of it and see if you're able to stick with it before you decide whether or not it's right for you.

Special Concerns for Women

If you're a woman of child bearing age, you'll need to make sure you are not pregnant before or during the hCG program. The hCG hormone is an indicator of pregnancy and plays an important role. You should not combine this treatment with pregnancy.

In addition, if you are trying to get pregnant or on fertility treatments you should also avoid the hCG diet. Not only is it unsafe for pregnant women, but the hCG hormone injections or drops can actually cause you to have a false positive on a pregnancy test.

If you're on hormone replacement therapy due to menopause – either brought on by a hysterectomy or age – you should not take part in the hCG diet. Adding more artificial hormones to your regimen is not advised.

Women who are taking birth control can participate in the hCG diet. While hCG is a hormone produced in higher quantities during pregnancy, it will not decrease the effectiveness of oral or other hormonal contraceptives.

Diabetes

There are some health conditions that make the hCG diet either difficult or dangerous to follow. One such case is diabetes disease. If you are diabetic, you may be able to follow the hCG plan, but you'll have to carefully monitor your blood sugar.

In this case, it's best to follow this program only under the care of a knowledgeable physician who can monitor your progress and make sure your health is not compromised. This is especially true because your medications may need to be altered.

That said, if you have type II diabetes, weight loss can actually help to reduce the effects of the disease. You may find that following the hCG diet allows your health to become better. In any case, make sure you consult your doctor before beginning this program.

Cancer

If you've been diagnosed with cancer or are a survivor of cancer in remission, you should not take part in the hCG diet. Hormones are often linked to the growth of tumors and cancerous cells.

If you have a history of cancer, you should not add artificial hormones to your body. The hCG diet is not advisable unless you have spoken with your physician and he or she has approved the treatment.

You should also not follow the hCG diet if you have any kind of tumor or abnormal growth that hasn't been diagnosed. Hormones can actually increase the growth rate of tumors and other growths and following this plan could worsen your condition.

Thyroid Problems

If you have thyroid problems, this is another case where you will want to talk to your doctor before making a decision. Some people with thyroid disorders can follow the hCG diet as long as they are taking medication and are monitored.

For other people, the thyroid problem rules out the hCG diet as a weight loss plan. If you're being treated for hypothyroidism or hyperthyroidism, make sure you talk with your healthcare provider before beginning the hCG protocol.

Breastfeeding

If you have recently had a baby and are breastfeeding, you'll need to talk with your doctor before beginning the hCG program. As this program is hormonal, you'll need to make sure that it won't interfere with your milk production or the health of your baby.

Budgetary Concerns

Most people would say they can't put a price on their health and happiness. However, the reality is that most people have limited funds to spend. When it comes to the hCG diet, you'll want to consider if you can financially budget for the program.

Some of the cost associated with the hCG diet include the cost of the injections or the homeopathic drops, physican visits, lab work, and purchasing special grocery items during the diet phase.

While some of these costs may be covered by medical insurance, the hCG diet is not covered by all insurance plans. If yours does cover it, you can expect to pay a copay for office visits and medications.

If your insurance doesn't cover it, you'll need to shop around for the best prices of the hormone. You can purchase homeopathic drops of hCG online and at health food stores. You'll need to comparison shop to find the best deal.

Even if you go with prescription injections, you'll find that the price of the hormone varies widely from pharmacy to pharmacy. Call around to see who has the best price and to determine if it's in your budget.

Are You Ready for a Change?

While many people say they want to lose weight, not every person is ready to take the necessary steps. You're the only one who can know if you're ready to lose weight or not. How long have you been talking about losing weight? Have you tried other programs?

These are questions to ask yourself when you're considering whether or not to begin the hCG diet. Once you begin this diet, you'll be restricted to a small number of calories for a short period of time. But if you stick with it, the long-term results are worthwhile.

Before you make a decision, here are a few ideas to help you make an informed one:

· Talk with your physician and make sure you are healthy enough to participate in the hCG diet.
· Find a physician that administers the program and assess the financial cost.
· Locate resources for homeopathic drops and assess their cost.
· Talk to other people who have tried the hCG diet and find out about their experiences.

Only after you've investigated the hCG diet and taken a look at your specific situation can you make a decision. In the next chapter we'll discuss the possible side effects of following the program. This chapter will also help you to make a decision.

Does the HCG Diet Have Side Effects?

What to Watch for When You Begin

Any time you begin a new exercise program that involves taking pills or adding supplements to your diet, you're naturally curious about side effects. In many cases, diet supplements have been found to cause major health problems.

However, with the hCG diet, you'll worry a lot less. The hormone being added to your routine may mean you feel some changes, but you won't need to worry about some of the scary side effects you get with diet drugs.

The Danger of Diet Drugs

Before we talk too much about side effects, it's important to remember that hCG is not a "diet drug". Instead, it's a hormone that occurs naturally in the body. In the case of the hCG diet we're harnessing what nature has given to help burn fat.

But with most diet remedies on the market, the same can't be said. In fact, many diet medications can be extremely dangerous and end up being pulled by the FDA because of side effects.

Side effects of diet drugs can include:

- Increased heart rate
- Increased blood pressure
- Nervousness

- Dizziness
- Cramping
- Diarrhea
- Gas
- Headache
- Dry mouth
- Insomnia
- Constipation
- Increased risk of heart attack

With all of those problems, it's no wonder people have concerns when starting a program involving supplements. However, you can avoid these dangerous side effects by following the hCG diet.

What to Expect with hCG

When it comes to the hCG diet, you'll be adding either hormone drops or hormone injections to your body. This hormone may cause you to have some slight noticeable changes, but most of these go away soon after you begin the program.

Some people report experiencing mild headaches and lower back ache. However, these symptoms begin to disappear as the body gets accustomed to the change in hormonal levels. Women also sometimes report feeling breast tenderness. This, too, subsides quickly.

In addition to breast tenderness, a sudden change in hormone balance can also cause some changes in mood swings and sex drive. However, these tend to be mild and also subside as soon as the hCG supplement is stopped.

Toward the end of the treatment protocol, some people report feeling muscle weakness. This is due to the loss of fat around the muscles. The muscles become too long as fat disappears. As a result, they have to contract harder in order to do the same amount of work.

The good news is that this side effect disappears completely after the hCG treatment cycle is over. While it can be a strange feeling, it isn't life threatening and most people find it merely an inconvenience.

When an hCG cycle is close to ending, some patients also report having low blood sugar. Low blood sugar can make one feel weak and even a bit dizzy. However, this symptom also goes away when the maintenance phase begins.

If you begin to have problems with low blood sugar, you will need to increase your sugar intake. This should be monitored by a physician to make sure any adjustments to your diet program are made properly.

Most people who go on the hCG diet report few to no side effects. And the really good news is that none of the side effects from the hormone are life threatening. The most important thing to remember is that regaining your good health and an optimal weight outweigh any negatives.

If you begin to experience strange or unusual side effects when you begin the hCG diet, you'll want to talk to your healthcare provider. It's always better to be safe than sorry. As you lose weight and begin to pay more attention to your body, you may uncover an underlying health problem you didn't notice before.

A Short Course

Unlike other diet medications and supplements, the hCG diet is meant to be taken in short courses. You'll never take hCG hormone for more than six weeks at a time. And in between courses you'll wait for weeks or months.

Your body won't be exposed to hCG long-term as it would be with diet drugs. As a result, any side effects are mild and short-lived. You don't have to worry about the long-term exposure to stimulants and appetite suppressants that other medications give.

If You're Still Concerned

In spite of what you've read here, you may still have concerns about side effects from the hCG diet. It's perfectly understandable that you might have reservations – especially if you've tried diet medications in the past.

The best way to put your mind at ease is to find out as much as you can. Make an appointment to speak with your healthcare provider about the plan and how it might affect your body. Talk to other people who have gone on the hCG diet.

Make sure to ask the questions for which you really want answers. Find out first hand what it's like for someone beginning this weight loss journey. You are the best one to make a decision about what is right for you, and getting the most information you can is vital.

Making a Decision

Once you've asked all of your questions, gotten the okay from your doctor, and decided that this plan is within your means, it's time to make a decision about moving forward. It's time to get real about yourself and what's most important to you.

If you're looking for a plan that will allow you to achieve permanent weight loss without having to be on a diet forever, the hCG diet offers hope and help. This is a plan that you can follow short-term to get long-term results.

Losing weight is a personal decision. It's often as much about health as it is about fitting into new clothes. If you want to be able to move more easily, have a longer life, and feel the freedom of being able to run, jump, and take on the world, this plan can take you there.

The hCG diet's benefits far outweigh any of the mild side effects that you may experience while following the program. Instead, you'll experience the positive side effects of good health and boundless energy.

Exercise and the hCG Diet

How Exercise May Actually Hinder Progress

If you're like most people reading this book, you've tried every diet known to man. You've counted calories, fat grams, and carbs. You've given up entire food groups and you've focused your entire diet on one food group.

In addition to all of these changes, you've also been required to add exercise into your diet. You've joined gyms, purchased videos, and hired personal trainers to help you carve the physique you've always wanted.

Why Traditional Diet and Exercise Programs Fail

When many people begin diet programs, they are already feeling overweight and fatigued. The absolute last thing they want to do is start exercising. But after reading a book or watching a program on health they decide that this is the only way to drop the pounds.

So they go all in. They buy the gym membership and sign up for the personal training package. They work out 7 days a week while their trainer Barbie tells them to just do a few more reps. Feeling the burn means you're doing the work, so you end up so sore you can't move.

After a while, the real world gets in the way and you miss an appointment or two at the gym. Or you even get so burnt out from overdoing it the first few weeks that you give up completely – or even worse, get hurt.

Over time, you find that you're paying for a membership to a club you never go to anymore. Barbie is just a distant memory. And now you feel more like a failure than you ever have before. Cue eating too much and gaining more weight.

Does any of this sound familiar to you? If so, it's time to break that cycle and try a whole new approach. With the hCG diet, you won't be asked to jump onto the exercise roller coaster that some diets ask of you.

hCG is Different

There's nothing wrong with getting exercise – and it is good for you. What's not good is trying to overdo the exercise while you're restricting your calorie intake. This will actually slow down your weight loss while you're supplementing with hCG.

The only exercises that are recommended for those following the hCG protocol are walking, stretching, and yoga type exercises. And while adding these to your routine is acceptable, you shouldn't add more than 15 minutes daily.

During the hCG diet, you'll want to avoid any strenuous exercise such as heavy aerobics, weight training, or any other high intensity program. While your body is safely burning fat, you're still only consuming 500 additional calories each day.

This low calorie intake can't support you if you choose to exercise too vigorously. So while you're in the active Phase Two of the program, you need to back off of exercise and keep things light.

In fact, if you do exercise your body may switch from fat-burning to starvation mode. In this mode it will hold on to every bit of caloric intake it can to save for later. This is the opposite of what you need and want for your weight loss.

For many people, seeing a halt on weight loss causes a lack of motivation to continue the plan. Instead of trying to force something to happen, allow your weight loss to continue naturally and follow the advice to avoid strenuous exercise.

While this may seem difficult for some people, other people find it a welcome relief. The good news is that when you're in the maintenance phase, you'll be able to gradually add more exercise into your routine if you want to.

More Motivated During Maintenance

Once you've completed Phase Two and are on to the maintenance plan, you may return to the gym if you choose. Many people actually feel more motivated than ever because they've lost weight and exercise has become easier.

Instead of feeling heavy on their feet, people who have lost anywhere from 5-40 pounds feel lighter and have more energy than before. Going to exercise won't feel like as much of a chore as it once did.

Suggestions for Getting Started

If you do decide to get started with an exercise program, there are many options. Make sure to start out slowly. You've been unable to exercise for several weeks and your body has undergone a great change – for the better.

Don't try to go from zero to 60 with your exercise program. Start out slowly and you can work your way to more strenuous exercise as you become stronger and leaner. Exercise is also a great way to help maintain your weight loss when you're in the maintenance phase.

Here are a few ideas for getting started with exercise – especially if you've never been a big fan:

· Grab a buddy – people often have more success if they have an exercise partner

· Start with walking – it's cheap, flexible, easy, and you already know how to do it

· Focus on flexibility – try yoga or other stretching exercises to lengthen your muscles and relax your mind

· Add resistance – adding resistance to your muscles can help you to maintain strong bones and keep your metabolism revved up

· Use light weights – there's no need to look and feel like a bodybuilder, strive for toned and lean

· Go on an adventure – find something you really love to do like hiking, canoeing, or rollerblading. It will make exercise so much fun you wont' know you're doing it.

· Seek professional help – a trainer can help you to discover an exercise program you both enjoy and that helps you meet your goals

· Dress appropriately – make sure you have shoes in good condition, wear breathable clothes, and wear the proper support

If you've tried exercise before without success, the hCG diet will help you to have the energy you need to give it another shot. Having already lost weight, exercise will be easier and more fun.

Not to mention that exercise has some great health benefits – both related and unrelated to your weight – including:

- Maintaining a healthy weight
- More flexibility
- Fewer aches and pains
- A stronger heart and decreased risk of heart disease and heart attack
- Decreased risk of cancer
- Decreased risk of osteoporosis
- Ability to lift objects more easily in daily life
- Better posture
- Muscle tone
- More energy
- Improved sleep
- Improved mental health
- Stress reduction
- Anxiety reduction
- Improved skin tone
- Better digestion
- Ability to run and play with your kids
- And the list could go on and on and on....

While you may have had negative experiences with exercise in the past, the hCG diet offers you a chance to set things straight and start fresh. During phase two, remember that you shouldn't exercise strenuously.

As you're taking the hormone and eating only 500 calories each day, you should keep exercise limited to walking, stretching, and low impact activities. Fight the urge to go to the gym and exercise vigorously – it will only cause you to slow your weight loss.

Once you've completed the second phase and have moved to maintenance, start slowly. You'll have more energy than you did before and exercise will put less stress on your joints. This will motivate you to want to move more.

You may enjoy joining a gym or trying strenuous exercise, but if you don't that's okay, too. Stick to moderate activities such as walking and stretching and make those a part of your daily routine.

Living a Balanced Life

Most people who complete the hCG diet find that they naturally eat less and have more energy to go through their daily lives. Whether or not you choose to make exercise a central focus in your life, you'll experience these benefits.

With the hCG diet, you don't have to be afraid of exercise. Forget about your past experiences with exercise that caused you to feel miserable, in pain, and even made you feel like a failure. With the hCG diet you can have success while adding minimal exercise.

But when you do add exercise to your life you'll also add many health benefits that can't be attributed to weight loss alone. For the best life possible, you'll want to live a healthful, balanced life.

Journal Your Way to Success

Keeping Track of Your Accomplishments

When you're starting a weight loss journey, keeping track of your success is essential. There are several reasons for this. Keeping a journal can help you with many aspects of dieting that make it a great tool.

Journaling is even more critical when you're following the hCG diet. That's because when you are on this program, you need to weigh yourself daily. This will help you to see your success, but it will also help you to stay on the program.

Daily Weigh In

When you start this diet program, you'll be asked to weigh every day. This will help you to track your progress while you're following the second phase of the diet plan. Don't be surprised if you're losing as many as 2-3 pounds on this plan.

Some days you may only lose .5 pounds, and that's okay, too. As long as you're losing some weight you can be sure you're following the plan properly. There are a few times when you may see that your weight loss stalls.

For example, many people have a slow down in weight loss about a week into the program. This is normal and shouldn't be anything to worry about. For women, you may experience a slow down when you start your period.

As this program is based on hormones, changes in your body's hormones will also cause changes in the rate of weight loss. If you're following the program to the letter and you're experiencing no weight loss, it's usually a result of hormones.

In fact, women who experience a drop in weight loss or even a weight gain (as long as they're following the program perfectly) are advised to take a pregnancy test. This major change in hormones can cause the hCG diet plan to lose effect.

However, the pregnancy test will only be accurate if you stop using the hCG hormone for at least five days first. Remember, pregnancy tests are actually measuring hCG levels in the urine and if you're taking hCG you could get a false positive.

During the maintenance phase of your program, you'll use the daily weight in as a tool to determine how many starches and sugars you can add back into your diet. If you gain more than two pounds in a single day, you need to decrease those sugars.

Two pounds is not very much to measure, so a daily weigh in will help you to keep track of those fluctuations. Keeping a journal will help you to keep track of your daily weight and keep you up to date on your progress.

Keeping Track of Your Food

Most people really don't have a good idea of how much they eat or how many calories they consume each day. Using a journal will help you to really see with your own eyes what you're eating.

With the hCG diet, you're asked to restrict your calories to 500 per day. This is very little food. It helps to write down what you're eating to help stay honest about your adherence to the program.

Even eating 100 extra calories – 3 or 4 crackers worth – can throw you totally off track during phase two. Writing down every bite will help you to see what you're eating and help you to remember what you still need to eat for the day.

During the maintenance phase, you'll be able to add back all the foods you love. However, you'll need to do this a little at a time. Writing down what you eat will help you to see how many sugars you're adding each day.

Then when you have a weight gain, you can determine where you need to cut back. If you don' t have a record of what you've eaten, it will be hard to adjust your diet plan to keep the maximum benefits of your initial weight loss.

Emotions Eating You

While we like to blame our weight on bad genes, a sedentary job, or love of food the truth is that many people who are overweight suffer from problems with emotional eating. Keeping a journal can also help you to combat these problems.

When you get a handle on your emotions and stop using food as a way to self-medicate, you'll find that your weight loss is permanent – and even something you don't have to think about anymore.

Keeping a journal of how you feel every time you eat can help you to see what triggers your problematic eating. Here are a few emotions you may uncover:

- Happiness
- Sadness
- Loneliness
- Frustration
- Boredom
- Anxiety
- Stress
- Anger
- Love

In the end, just about any emotion can lead you down the path of overeating. Keeping a journal can also help you to stay mindful about how full you feel. Many people eat until they feel like they are going to pop.

The result is a feeling of misery, indigestion, and weight gain. You may find yourself saying that you'll never make yourself feel that way again. Then, something triggers you to eat again and you repeat the pattern.

Instead, it's good to eat until you feel satisfied. You won't feel stuffed or full. You simply won't feel that your stomach is hungry anymore. It takes a lot of practice to learn how to eat in this healthy way if you're used to binging.

A journal can also help you to write about your feelings and get to the bottom of what's causing your food addiction issues. Perhaps you have had some experiences that have led you down the path of emotional eating. Journaling can help you to become proactive and deal with your problem.

Using the Past to Propel You Forward

Everyone experiences hard times now and then. When it comes to dieting, you're bound to have times when you experience great success and then times you feel like a failure. Being able to look back on good experiences can help you to take those tools and get you back on track.

When you don't write things down, it's hard to remember exactly how you solved a problem, how you felt when you lost weight, what tricks helped you to stick to the program, and what motivated you to start the hCG diet in the first place.

A journal is a record of your history. It can record your highs and lows. It will be a volume you can go back to when you need to seek encouragement from none other than yourself. It can also be a tool you use to help others when they need it.

Choosing a Journaling Method

In modern times, there are many ways to keep a journal of your diet journey. You'll want to choose a method that fits into your lifestyle and that you will stick with. A journal doesn't do any good if it's not something you can access easily and comfortably.

Paper and Pencil

An old fashioned blank book can be a fabulous way to keep a journal. You may be very comfortable with the type of gift books you can buy at the bookstore or even a spiral notebook that's from an office supply store.

Paper journals can be small enough to travel with you and they won't ever crash the way a digital storage device can. They can also be placed on a bookshelf where you can look upon the volumes of your life with satisfaction.

Date Book

A calendar or date book can make an excellent journal. It already has the pages dated for you, all you need to add is the essential information. Make sure to choose a calendar or date book that has plenty of room to write.

Digital Diary

There are many software programs available to help you keep a diary on your computer. Many are free and can be stored on your hard drive so that only you can have access to them. You can even password protect your journal if you choose.

Blog

One of the most popular ways to record the events of ones life in the modern age is the blog – or internet journal. There are many sites that host free blogs where you can personalize your page with photographs, pictures, and videos.

With a blog, just be advised that a free blog site can delete your blog at any time. You may want to back up your information so that it doesn't get lost forever. You can also decide whether or not to make your blog public. Many people like to share support through blogs.

Social Networking Site

Some people prefer to use a social networking site to share details of their diet and weight loss journey. Sparkpeople.com is one of the most common sites for people working on weight loss and health goals. You may want to look into setting up a free profile.

This site will also give you the ability to share information with other people on a similar journey. You can get great motivation and support from online communities like this and many others.

How Much Is Enough Weight to Lose?

Setting the Right Goal Weight

When you set out on a diet, you should begin with the end in mind. But how do you know what an appropriate weight is for your body? In this chapter you'll learn how to determine the best goal weight for you.

You need to strike a balance between being optimistic, realistic, and most importantly healthy. Many people have no idea where to even start. But it's a good idea to begin with taking a look at your history.

Your Body's History

Many people who weight over 200 pounds dream of being 115 pounds. That may be a realistic goal, but it may not. The first thing to do is take a look at your body's timeline. The easiest way to do this is to get out the family photos.

Make a timeline of yourself from the time you were a very small child until the present day. It doesn't have to be perfectly accurate, but do your best to come up with accurate dates to put the photographs in order.

If you can remember, make a note of the weight that you are in each picture. If you can't remember, that's okay, too. You can probably estimate just by looking at your pictures. The main thing you need to do right now is establish a pattern.

Have you been heavy ever since you were a small child? Did you gain weight later in life? This information will help you to go about setting a realistic goal for your body. It's not that you can't weight whatever you dream of, it's just that you need to be clear about your own body.

If you've never been a size 6, it may be that your body wasn't meant to be a size 6. For some people, it's perfectly healthy to be a size 12. In fact, for some people a size 12 is healthier than a size 6 could ever be.

BMI

Another number to take a look at is your BMI, or body mass index. This is a number that health professionals use to determine how much you should way based on your height. It's actually a complicated formula turned into something simple.

One thing to keep in mind is that the BMI is not a perfect number. Many people shoot for unrealistic goal weights based on the BMI. It should only be used as a guideline to help you determine a goal weight, but it's not set in stone.

In general a BMI score of 18.5-25 is considered a healthy weight based on your height. From 25-30 is considered overweight, and above 30 is considered obese. Below 18.5 is considered underweight.

There are many online BMI calculators you can use for free. They'll ask you to input your height and weight and then they will crunch out your BMI. In addition, they'll also tell you what your ideal weight range is. You can add 10% to the top number and still not be overweight.

Remember to take this number with a grain of salt, however. While it's the standard used in medical practices it isn't perfect. You'll also need to take into account your body type and your age.

Once you've determined the BMI number that's appropriate for your height and age, you can use that number as a guideline to help you make a goal weight. Remember, though, that this number isn't perfect. In fact, for some people the BMI is much too low for their actual body.

Take the Plan into Account

With the hCG diet, you also need to think about the limitations of the actual plan. For example, you may have 100 pounds to lose, but you can only lose up to 40 pounds in a single cycle of the diet.

Instead of trying to set a goal to lose 100 pounds, you might want to start smaller. It's also important to remember that some people only lose half a pound a day on the diet. That means that while many people can lose 40 or more pounds the first cycle, not everyone does.

With the hCG diet, you also need to remember that you have to follow the plan to the letter. If you don't, you won't have the kind of weight loss results you seek. But if you're losing at a slow rate even following the plan perfectly, there's not much you can do to speed things up.

You certainly can't eat fewer calories or exercise with the program. In other words, you may have to be more patient than you'd like while you're following the hCG diet. In fact, before you set a short term weight loss goal, you may want to go ahead with the first cycle.

This way you can see how your body will respond the course of hormone and calorie restriction. You can use this as a benchmark to set goals for the rest of your diet program as you lose until you reach your final goal.

Don't Set the Bar Too Low

If your goal is to weight 110 pounds, but you haven't weighed that since junior high, you may need to rethink your goal. Some people set weight loss goals that are just simply unattainable for them.

You'll have more success if you have a realistic goal in mind. Now that you know your BMI, go back to your timeline and take a look at your previous weights. When were you last happy with your weight?

Perhaps growing up you always thought you were overweight, but as an adult you look back and realize that you really weren't. Try to set a goal that is something you can realistically attain. Don't try to be a supermodel weight.

In today's media, it seems that everyone is very thin and that it's more natural to have hollowed cheeks and sticks for arms and legs. But the truth is only 2% of the population is naturally that thin. The rest of us need to shoot for a goal that's healthy and a little thicker.

Size vs. Weight

Weight is also not the only way to set a goal for yourself. For many people, it's just as important to attain a specific size. And it's easier to determine if you're losing the weight you want to lose when you see how your clothes fit.

Consider setting a goal that's related more to your size than the numbers on the scale. Those numbers will fluctuate rapidly and don't tell the whole story. Your size is a better indicator of whether or not you're having success.

You can also measure the inches on your body to determine how much you're losing. This is a great indicator of how well you're doing. The following areas should be measured when you're keeping track of your inches:

- Upper arm (widest part)
- Thigh (widest part)
- Hips
- Waist
- Bust/Chest

It helps to have someone else take the measurements. Make sure they take them at the same exact spot each time. A weekly measurement will give you a good indicator of what's going on with your weight loss.

Keep track of your inches in your weight loss journal. Many people enjoy entering them into a spreadsheet program where you can graph data. This way you can literally watch your weight and size decrease.

Setting your goal weight is an important step in following the hCG diet. It will help you to keep your focus on the program. It will also help you determine how many cycles of the plan you need to follow.

Set Small Goals

You'll want to have one big overall weight loss goal, but it also helps to have smaller goals for your weight loss. For example, during the first cycle you may want to have the goal of losing 20 pounds. You may lose more, but this is a good starting point.

If you set small goals and have success, you'll find that it makes it much easier to keep going with the program. It's a snowball effect that can help you to reach the bigger goal you've set for yourself.

Set Goals Not Related to Weight

Then you can set goals that are not related to weight at all. For example, you can set a goal of taking a 15-minute walk each day while you're on the hCG diet. This is a good way to begin the habit of exercising without overexerting yourself.

The good thing about setting a goal not related to weight loss is that you are absolutely in control of whether or not you achieve it. Your weight is not something you are directly in control of – there are so many factors that play into it.

Here are a few goals you may want to think about adding to your list.

I will:

- take my hormone each day at the correct time and dose
- follow the eating plan perfectly – no more than 500 calories each day
- try relaxation and stretching exercises
- write down everything you eat each day
- weight daily
- report my progress daily to a buddy
- keep track of my emotions surrounding eating
- keep up with any doctor appointments and lab work necessary

When you set goals that aren't related to weight loss, you absolutely can attain them in spite of your body's metabolism or cooperation. With the hCG diet program you'll find that setting this type of goal will help to keep you motivated.

It Takes Time

While the hCG diet is a fast weight loss program, you need to remember that big weight loss will still take patience. You have to space out your cycles of the hCG diet. Each time you start a new cycle you have to wait longer between cycles.

If you have a substantial amount of weight to lose, it can still take a year to lose 100 pounds – even if you lose 2 pounds a day when you're on the hormone and calorie restriction phase. You can't allow yourself to become discouraged.

In between cycles, you'll still need to make sure you're eating sensibly, exercising moderately, and getting prepared for the next cycle of the diet program. While the hCG diet is a tried and true method of weight loss, it may still require you to be patient.

Reward Yourself for Reaching Goals

When you do reach goals – whether they are related to weight or not – you need to take time to celebrate your success. With the hCG diet program, it's not a good idea to use food as a reward, but there are many other things you can use as a reward. For example:

- treat yourself to a new item of clothing as you're losing weight
- buy a piece of jewelry you've been wanting
- get a massage
- go to a movie with a friend
- get a manicure
- take a mini-vacation about which you've been dreaming

When you allow yourself to celebrate small successes, you'll help yourself to stay motivated to focus on the plan. The hCG diet plan can be very restrictive, so it's important that you give yourself credit for following it.

When you reach you final weight loss goal, you'll have a lot to celebrate. Start planning now for how you will celebrate that goal. Maybe you will start saving now to buy some new wardrobe pieces, take a vacation, or take a new adventure.

Get Ready, Get Set, Go!

When you're starting any new lifestyle change, it's important that you prepare ahead of time. Not being prepared and organized is the number one reason why people fail at a diet program. If you're rushing around and don't have what you need, you won't have success.

For this diet, there are some particular supplies you'll need to get started and to maintain your loss. In this chapter we'll go over what you need to have success and provide you with an easy checklist for getting started.

There are also actions that you need to take not directly related to supplies. But these are just as important as having the right items available to you. We'll also take a look at what action steps you need to take.

Check with Your Healthcare Provider

The very first thing you need to do before beginning the hCG diet is to check with your healthcare provider and get the okay. You need to make sure you don't have any underlying issues such as diabetes or heart disease before getting started.

For women, you also need to make sure that you are not pregnant or trying to become pregnant before beginning the hCG diet. It's also a bad idea to begin the hCG diet if you're being treated for infertility.

Before beginning any weight loss program, it's a good idea to get the all clear from your doctor. With the hCG diet, you may also want to go to your doctor for monitoring and to get prescriptions for the injections.

Make an appointment to discuss the options with your healthcare provider before beginning. This way you can make sure you follow the diet safely and successfully. Your doctor can be your partner in your weight loss program.

hCG Supply

When you begin the hCG diet, you'll need to follow it to the absolute letter. That means you can't miss even a single dose of the hormone. It's critical that you have an adequate supply before you get started.

If you're taking injections, you'll have to have a prescription for the hormone. If you're choosing to use homeopathic drops, you can buy them without a prescription. Many companies make starter kits that have a complete course of the hCG drops.

This is a good idea before you get started. You'll insure that you won't run out of hormone before you finish the cycle. If you don't want to purchase a starter kit, calculate how much you'll need and purchase a complete cycle.

This is the most expensive part of the program, so be prepared to invest in your dieting success. If you're going the homeopathic route, you'll spend less than you will on injections. It's a good idea to find out the differences in cost and plan accordingly.

If you're using injections, you'll also need sterile alcohol wipes, syringes, and a sharps container for disposal. You may also want to get a few extra syringes for practicing injections. Many people practice on an orange to get the hang of it. Your doctor will help you to know what to do.

Filling Your Fridge

You'll also need to have a supply of fresh foods to eat while you're following this diet plan. While you're restricted to 500 calories each day, you also need to make sure you're eating the right foods with those calories.

Because most of the foods you eat will need to be fresh fruits and vegetables, you'll find that you have to make more trips to the supermarket than you may have made in the past. Schedule a day that you will shop each week.

Create a grocery list that you can use each week. One of the easiest things to do is to plan a weekly menu and then just repeat it each week until the hCG cycle is over. This will make it convenient to shop each week.

Your list will need to include:

- fresh fruit
- lemons
- fresh vegetables
- lean meat such as chicken and fish
- melba toast
- grissino bread stick
- coffee or tea
- fat free yogurt
- fat free cottage cheese

Spice It Up

With only 500 calories each day, you may find yourself getting pretty bored of your food. This is a good time to experiment with fresh herbs and spices. If you haven't been much of a cook before, you'll probably need to stock up on spices.

Some good spices to add to your pantry include:

- salt
- pepper
- cinnamon
- paprika
- lemon pepper
- season salt
- parsley
- Italian seasoning
- Poultry seasoning
- Steak seasoning

You'll also want to have fresh lemons on hand. With the hCG diet it's recommended that you consume the juice of one lemon each day. Bottled lemon juice often contains added preservatives or sugars – so make sure to use real lemons.

Finally, you'll also need to avoid sugar as a sweetener. Instead, you'll want to use stevia. This is a natural sugar that comes from a plant. It doesn't have any added chemicals and it doesn't have any calories.

Free Foods

With the hCG diet, you may want to take advantage of foods that are free and can be eaten at any time. These are foods that have no calories and won't add to your 500 calorie a day limitation. Having a few of these items in your pantry will help you when you're feeling like you need something extra.

Free foods include:

- Miracle Noodles – these are noodles that have no calories or fat and come in many varieties
- Chicken broth
- Beef broth
- Vegetable broth
- Sugar free jello
- Herbal tea
- Coffee – black
- Black tea
- Green tea

Eating Organic

While you're following the hCG diet, you need to stick to foods that are produced in the most natural way possible. As often as you can, purchase organic produce and meats. These will have fewer antibiotics, pesticides, and hormones.

You'll also want to take a look at the products you put on your skin as these become absorbed by the body. For example, while following the hCG diet you should avoid any lotion that's not produced naturally.

The basic rule of thumb is that you shouldn't put anything on your skin or hair that you wouldn't eat. While it's best to avoid lotions altogether, many people choose to continue using them. If you do, choose lotions and creams that are produced organically.

Purge Temptation

Chances are, your pantry is full of processed foods that include high amounts of sodium and sugar. If this is the case, you need to make sure that you aren't tempted to go off of the hCG diet with these items around.

The best thing to do is to empty your pantry of any foods that may cause you to stray from the hCG plan. If you only have the items that are sanctioned by the diet plan, it will be harder to go off of the diet.

If you don't want to get rid of these items permanently, box them up and send them to someone else's house for a time. Remember, though, that the once you've completed the hCG protocol, you'll probably have different tastes.

You'll want to begin to eat healthier foods that aren't processed as part of maintaining your weight loss and good health. You may find that these items never get an invitation to come back to your home.

Odds and Ends

You'll also need to have some basic supplies to help you with your diet efforts. These don't all fit into one category. Rather they're miscellaneous things that you need. They include:

- Scale
- Journal
- Food scale

- Large water cup/bottle
- Binder to organize supply lists and menus (you may also use this as a place to keep track of your weight and your food consumption)
- Tape measure for taking inch measurements
- Containers for taking food to work

Preparation Equals Success

Because you're only eating a small amount of food, you don't have a lot to purchase. But making sure you have the proper things will help you to have success you desire. Getting them before you even start will make it an easier transition to the new lifestyle.

When you're disorganized, it's difficult to stick to any type of routine. With the hCG diet, you'll need to be very self-disciplined. Rather than find yourself feeling chaotic and stressed out, it's best to plan ahead, get organized, and be prepared.

On the next three pages you'll find some very helpful resources to help you get organized. You'll find that keeping all of your information in one place is critical to your success. If you're seeking the assistance of a physician, a binder will be easy to transport to appointments.

You can keep your menus, grocery lists, track your weight, and keep any lab reports from the doctor in this binder. It makes it convenient to track your success when you keep everything in one place.

hCG Diet Preparation Checklist

☐ Make appointment with healthcare provider. Get the okay to get started with a new diet program, discuss options of hCG injections vs. homeopathic drops. Also verify that you are not pregnant, have basic blood work done to rule out diabetes or thyroid problems.

☐ Purchase hCG supplies. Either homeopathic drops or vials of hCG with prescription at pharmacy.

☐ If using injectable hCG, purchase syringes, alcohol prep wipes, and sharps container for disposing of syringes.

☐ Make weekly menu following 500 calorie plan

☐ Make grocery list containing all foods included in menu.

☐ Purchase scale if necessary

☐ Purchase food scale if necessary

☐ Inventory spices and add necessary spices to grocery list

☐ Determine what system you will use to track your weight, food intake, and goals

☐ Purchase necessary journaling supplies – book, pencils, software, etc.

☐ Empty refrigerator and cabinets of all foods that are not allowed on hCG diet

☐ Purchase containers for taking food to work – disposable containers or reusable containers, lunch sac, etc.

☐ Purchase large container for water – either a bottle or large cup. Remember you need 100 ounces each day!

☐ Set date to begin. For women, make sure you begin right after your menstrual period ends.

☐ Put together a binder that contains all the important information for your diet. You can also use this to store your favorite food lists and recipes.

☐ Make sure to stock up on free foods so that you always have them available when the 500 calories are not satisfying your hunger.

Weekly Menu Worksheet

Daily Guidelines:

☐ Vegetables – 1 serving, 2X a day

☐ Fruits – 1 serving, 2X a day

☐ 3 oz. lean meat, 2X a day

☐ 2 melba toasts or 2 grissino breadsticks – daily

☐ The juice of one lemon each day

☐ Fresh herbs and seasonings - unlimited

hCG Diet Daily Journal

Day:_____**Date:**_____

Daily Food Consumption

Breakfast	
Lunch	

Dinner	
Snacks	

☐ **hCG taken**
☐ **100 ounces water**

Weight:_____

Pounds lost/gained:_____

Total lost to date:_____

The Dos and Don'ts of the hCG Diet

When following the hCG diet, there are many things you can do to have success. There are also many mistakes that people make along the way. In this chapter we'll discuss the dos and don'ts of the hCG diet so that you can get the most success possible.

Do know what you're getting yourself into.

Talk with your healthcare provider and get as much information as you can ahead of time. The more you know, the easier it will be to stick to the plan and lose the weight. Read books, look at message boards, and get yourself into the mindset you need to follow the program.

Don't stray from the plan.

You can only eat 500 calories each day on the hCG program. If you stray from the program you'll find that you don't lose weight as fast as you'd like to. Even one day of eating too many calories will hinder your success.

As you can imagine, 500 calories goes quickly. However, most people on the hCG diet report that they don't feel hungry after the first couple of days and that they have to force themselves to eat the food each day.

Do take your hormone at the same time each day.

Don't forget to take the hCG every single day at about the same time. This will help your metabolism to get revved up and stay up. When you miss a dose, you'll see it on the scale. If you're someone who is very forgetful, you'll need to come up with a system.

Perhaps you can set an alarm each day to help remind you to take your hormone. You can also add it to something you already do normally – such as brushing your teeth or taking your shower. Once you get in the habit, it'll be easier to maintain.

Don't Exercise Strenuously

Again, it's important to remember that this isn't your typical diet and exercise plan. If you exercise vigorously, you'll actually prevent the diet from doing its job. It's okay to exercise moderately by walking, doing yoga, or stretching.

When you're in the maintenance phase you can go back to a more vigorous workout routine if you choose to. But while you're in the initial phase of the diet, don't exercise more than 15 minutes each day.

Do Seek Support

Many people are using the hCG diet to lose weight. When you're struggling with sticking to the program or if you have questions, it helps to have someone to talk with. If you don't know anyone in your immediate circle of friends who is trying the hCG diet, there are online resources.

There are many message board and social networking sites dedicated to people who are dieting and working to lose weight. A quick search online will pull up hundreds of resources. You'll find you can ask questions and get advice from people who are in the same situation.

Don't Start the Next Cycle Too Soon

If you've finished the first hCG cycle and had success, you're probably anxious to start the next cycle and lose even more weight. You may be tempted to start the next cycle right away so that you can get that weight off faster.

However, when you go back to the next cycle too soon you actually decrease the effectiveness of the program. Your body begins to become tolerant to hCG and your metabolism begins to slow back down.

When you want to have permanent weight loss on the hCG diet, you have to follow the protocol to the letter. Don't try to rush things even if you're very tempted. This won't help you to lose more weight – it will keep the diet from being effective.

Do Listen to Your Body

As your body is changing through the hCG diet, you may find that you notice those changes more. Dieting can cause you to become more in tune with your body and this is a good thing – if you listen to it.

While you don't need to rush to the doctor every time you have a little twinge or change, it's a good idea to listen to your body's signals. If you're having frequent headaches, fatigue, dizziness or other strange symptoms, make sure to get them checked.

Changing your hormones and eating a small number of calories can sometimes work against people who have conditions like diabetes and hypothyroidism. You may uncover an underlying health condition you didn't know you had.

Make sure that you listen to your body and take what you observe seriously. There's nothing wrong with getting your healthcare provider's opinion. He or she may order lab work to check out what's going on.

Don't Buy hCG from Just Anyone

There are many places where you can buy hCG – both injectable and in drops – online. But not all of those sources are reputable. In fact, unless you follow certain precautions you may not be getting something safe or effective. Here are some guidelines for purchasing hCG:

· Injectable hCG should never be purchased online unless it's from a highly trusted resource. You should always purchase it from a pharmacy using a prescription.
· Buy hCG that's manufactured in the United States and FDA approved.
· Pay attention to the price – if it's much cheaper than its competitors, there's probably something wrong.
· Make sure you buy with a money-back guarantee.
Remember that if something seems too good to be true, it probably be is. Beware of people who want to take your money and prey on your desire to lose weight. There are many reputable places to purchase hCG – use those!

Do Get Into a Routine

Changing your eating habits can be very difficult. One way to make things easy is to stick to a routine. Many people find that eating the same thing for breakfast and lunch and only having variety at dinner is easier than trying to mix it up all the time.

While it seems like variety makes things more interesting, it also makes it harder to stick to a routine. You only have to stick to 500 calories a day for a short time – don't worry too much about getting bored with your food choices.

Don't Stick to Bland Foods

When you're limited to a small number of calories and food choices each day, you want them to taste good. Don't be afraid to add spices and fresh herbs to make the food you're eating more palatable.

If you don't like the food on the diet program, you won't stick to it - even if you have to do it for only a few weeks. Bad food is bad food. Instead, focus on finding ways to add flavor to your food. Adding new spices to your cupboard is a great help in fighting the bad food blues.

Do Pack Your Lunch

Work is one of the hardest places to stick to a diet. There are people all around you eating foods that you can't have. There's also always someone bringing in treats for the break room. Be prepared to fight these temptations by bringing in your own food.

Pack your lunch every day so that you have what you need to get through your program on the hCG diet. Remember that when Phase Two is over, you'll be able to gradually add treats back into your diet. Be patient!

Don't Be Afraid to Ask Questions

When you're starting a new diet program, you may feel like you have a lot of information and become overwhelmed. Once you get into it for a few days, you'll start to have questions you didn't think about before.

Don't be afraid to talk with someone also following the plan or with your healthcare provider. There's no such thing as a stupid question – especially when it comes to your health and your diet.

Do Keep Your Journal

One of the best ways to stay motivated is to keep track of your success. While it may seem simple and mundane, keeping a journal can be the key to your weight loss now and the key to maintaining your loss.

Write everything down – your food intake, your daily weigh-ins, and your measurements. Also pay attention to your emotions. You can use your journal to help look back on successes and also to try to pinpoint problems that caused your weight loss to slow.

Don't Give Up!

The most important advice you can follow is to stick with the program. Weight loss can often be a difficult battle. If you're focused on your goals and celebrate your successes, it will help you to stay motivated.

When you give up, you often will revert back to old habits that caused you to gain weight in the first place. You'll give up the benefits that come from losing weight and keeping it off such as improved health and vitality.

The hCG diet is an important program for losing weight and keeping it off permanently. You'll be able to set a new pace for your metabolism and enjoy a lifetime of feeling great and feeling great about yourself.

While this diet has many restrictions at first, they don't last forever. When you start to feel overwhelmed, bored, or frustrated by the diet remember that it is a temporary phase. Within a short period of time you'll be able to add more calories and foods you enjoy to your plate.

Desserts for Desperate Moments

Indulging on the hCG Diet

When you're following a diet plan, you may begin to miss your old favorite treats. And while the hCG diet allows you to eat only 500 calories each day, there may be times when you feel desperate for a dessert or other indulgence.

This chapter includes some snacks that are part of the program. You can use them as part of your 500 calories or if you are desperate, you can have a few extra calories for the day. But the best thing to do is stay under 500 calories.

First Try Free Foods

Before you go off of the plan, try having some free foods to see if that curbs your hunger. For example, try a little bit of sugar-free jello and see if that quenches your craving for sweets. You can also try a bowl of broth with vegetables to fill you up.

Also try eating a piece of fruit or some vegetables to help give you that crunchy or sweet feeling you're craving. You may even want to start having a small snack between every meal to help curb your hunger.

After trying free foods, wait for at least a half hour before deciding you want to have a treat. This will give your brain time to realize you've put some food into your stomach. If after 30 minutes you still want more, try these treats.

Cinnamon Apple Crunch

This treat doesn't actually take you off of the plan – but it will make you feel like it did. Slice one apple and toss it with ¼ teaspoon cinnamon and a packet of stevia. Place them on a plate or in a bowl (microwaveable).

Then crush up one or two pieces of melba toast and sprinkle them on top. Microwave for about 2 minutes on high (depending on your microwave). Then enjoy! This is a tasty dessert that has all the spiciness and sweetness you expect from a dessert.

Fruit Smoothie

Try slicing up your favorite fruits – for example ½ an orange and 6 strawberries. Mix them with crushed ice and a packet of stevia. Add them to the blender and puree them – you may need to add a little orange juice to thin the texture.

Berry Sorbet

Choose a handful any berries you prefer. Strawberries and raspberries work well. Add the juice of one lemon (you need it anyway, after all). Add stevia and blend. You may want to add a little water to smooth the texture. Then freeze the mixture, stirring about every 30 minutes.

When the sorbet is the texture you enjoy, put it in a container that is resealable. You may want to make this up ahead of time to have as a treat whenever you like.

Cinnamon Toast

Take two pieces of melba toast and sprinkle with stevia and cinnamon. Broil them in the oven or microwave to heat.

Flavored Stevia Drops $8.99 @ Amazon

Flavored stevia drops are a great way to add flavor to foods without going off of the program. You can add these to carbonated water and actually make sodas that satisfy your desire to have something fizzy to drink.

You can also add flavored stevia drops to smoothies and desserts to get the flavors you miss without the extra calories. They can even be added to coffee in place of flavored creamers to give you that familiar taste.

Popsicles

Another tried and true treat for the hCG diet is Popsicles. You can make them with fresh fruit or stevia flavored coffee. Simply prepare your liquid, place it in a Popsicle mold – or ice cube tray with toothpicks – and freeze.

Focus on Your Goal

While it can be difficult to feel the cravings for sweet treats, remember that Phase Two only lasts for a short time and then you can add your favorite treats back to your diet in moderation. These suggestions can give you a little bit of indulgence without taking you off of the plan.

Using fruit and stevia, you can create a lot of delicious mixtures. Don't forget about spices such as cinnamon, ginger, and nutmeg to help create indulgent desserts. As you experiment, you may find some desserts that you'll continue to eat long after Phase Two is complete.

Maintaining Your Momentum

Powering Through Plateaus

One of the most discouraging things about dieting can be experiencing plateaus. This is a time that you stop losing weight. With any diet, even the hCG program, you may find that you experience plateaus.

Plateaus are a natural part of weight loss and if you know how to handle them, you'll power through and keep losing weight. These short breaks in weight loss don't mean the end of your program.

There are many things that can cause you to have a plateau. Sometimes knowing the cause can help you to fix the problem. In this chapter we'll look at the causes of some plateaus and how you can get back on track.

Exercise

With most diets exercising improves your results. And even though you've learned that isn't the case with the hCG diet, you may insist on trying to exercise. If you are exercising and have experienced a plateau, you need to back off of the program.

You should be doing no more than 15 minutes of very moderate exercise such as walking or yoga during Phase Two of the program. When you switch to maintenance and begin eating more calories, you can begin to participate in more strenuous exercise.

Too Many Calories

It's possible that you are following the program, but you are still actually eating too many calories. Take the time to write down everything you're eating. If you hit a plateau, you can go back and check to see if you have eaten too many calories.

Five hundred calories isn't very many. It doesn't take much to meet that quota before you are over doing it. Most people don't feel hungry even with such a small number of calories, but if you experience a plateau, check your caloric intake.

Too Few Calories

For some people, the plateau can come from not eating enough calories. You may be feeling very full because your body is using stored fat to burn calories. It can be hard to eat when you're not hungry.

But for the hCG diet to work, you need to eat at least 500 calories each day to make sure you keep your weight loss going. Look at your journal to see if you're following the program correctly.

If you're eating too few calories, try adding a serving of lean meat or another food that's dense in calories. It can be hard to eat when you're not hungry, but in order to keep burning the fat you'll need to eat the minimum number of calories.

Hormonal Changes

The science behind the hCG diet is primarily based on controlling hormones in your body to burn fat. That said, you may experience some plateaus that are simply do to hormonal fluctuations that are natural and normal.

Of course you should check your journal to make sure you are adhering to the program. But if you are and you're still experiencing a plateau just keep going. There may be a few days scattered throughout Phase Two where you don't lose any weight or you even put on a pound.

This is not a reason to panic. For women this can be particularly true. Women's menstrual cycles can often affect weight gain and loss while following the hCG diet. The best thing to do is stick with the program.

If a week goes by and you're still experiencing a plateau, you may want to speak with your healthcare provider to see if there's an underlying cause such as thyroid disorder. In most cases, your plateau will end and you'll be back to losing weight quickly.

Remember Your Purpose

When you experience a plateau, it can cause you to feel like a failure. Sometimes these feelings lead you to cheat or even completely give up on your diet. Do anything you can to prevent that from happening.

Weight loss is a hard business. It requires you to be committed, change your patterns, and be extremely regimented. But with the hCG diet, you only have to commit to do that for 3-6 weeks at a time.

Once the second phase of the program is over, you can begin to loosen up a little bit. You can gradually add back the foods you love and continue to keep the weight off. Don't let a little plateau keep you from reaching your goal.

It doesn't matter if you lose weight every single day. What matters is the pattern over a long period of time. If you're continuing to keep your overall weight loss total up and not gaining weight back, you're a success!

Beyond the Scale

The Health Benefits of Taking Off the Extra Pounds

Many people choose to lose weight because they don't like their size or they wish to wear different clothing. But for many people, the benefits of weight loss have much more important consequences.

There are many health benefits to losing weight. The hCG diet is the perfect way to melt away the excess pounds and improve your health. In this chapter you'll learn about how losing weight can help to secure a healthier future for you and for your family.

Energy

Many people who are overweight describe themselves as being tired. They often think that there's something wrong with them beyond just being overweight. But the truth is having excess weight can make you feel very tired and fatigued.

By following the hCG diet, you'll be able to shed excess pounds. And with every pound that comes off, you'll feel a little boost of energy. If you're looking for a way to have more energy to advance your career, play with your children (or grandchildren), or enjoy the social life you once had, this is the plan that will help you to do that.

It's simple, but increasing the energy you have will enhance your life in many ways. You'll feel like exercising more – and that has a whole set of health benefits in addition to weight loss. You'll also feel like you can face whatever challenge comes your way.

Freedom from Joint Pain

Another common complaint of people who have weight to lose is pain. You may have a sore back, aching feet, creaking knees, or all of the above. When your joints are under too much stress from weight, they tend to become damaged and in pain.

The good news is that when you take off the weight, you ease the pain of your joints. In fact many people who once thought they would be in chronic pain the rest of their lives find themselves pain free.

Even just a small weight loss – about 10% of your total body weight – can make huge changes in the way you feel. If you're tired of being in pain, the hCG diet can help to take your pain away and keep it away for good.

You'll also save money by not having to buy over the counter or prescription pain killers to ease your daily aches. It will make moving and exercising easier and you'll be able to participate in activities you have missed.

Lower Risk of Heart Disease

One of the greatest risk factors for heart disease is obesity. When you remove excess fat from your body, you take stress off of your heart. That immediately begins to reduce your risk of heart disease and heart attack.

You'll also find that losing weight can help to lower your blood pressure, cholesterol, and heart rate. These are major indicators of the health of your heart. With the hCG diet, you'll burn fat and calories and help to keep your heart in good repair. You'll also have more energy to exercise and keep your heart in shape.

Self-Esteem

While looking good isn't the most important thing in life, it can certainly help you to feel better about yourself. You'll also feel good knowing you can set a goal and accomplish it. All in all this will enhance your self-esteem.

When your self-esteem is improved, you'll find that you succeed in other areas of your life. Don't be surprised if you find yourself getting that promotion you've always wanted, starting to date someone special, or start a new hobby that's always held your interest.

It's not that losing weight has made you a better person; it's that the weight loss has helped to improve your confidence. When you have more confidence, you're able to accomplish a great many things.

Reduced Risk of Cancer

A direct correlation has been found between cancer and obesity. When you lose weight you can help to fight the genetic predisposition you may have for certain cancers. While weight loss isn't a guarantee that you won't get cancer, it certainly can help prevent it.

With the hCG diet you can move from being obese or overweight to being the weight that's healthy for you. You'll help to prevent cancers such as colon cancer that has been found to be strongly linked to weight.

Fertility

Many women experience problems with fertility. While there can be many reasons for this, weight has been found to be linked to fertility troubles. Women who lose as little as ten percent of their body weight can experience improved fertility.

While you don't want to participate in the hCG diet while actively trying to get pregnant, you can use it to lose weight in preparation for trying to have a baby. There's no greater reward for your weight loss efforts than being able to add to your family.

Decreased Diabetes

Type II diabetes is directly linked to obesity. People who are very overweight can have problems with insulin production and end up having to test their blood sugar and take medications to control the problem.

Another way to prevent and control diabetes is through weight loss. If you're prediabetic, following the hCG diet can keep you from ever becoming diabetic. It can help you to control your weight and prevent the disease.

If you are currently diabetic, you need to talk to your physician before beginning the hCG diet. With his or her approval, however, you may be able to participate in this plan to lose weight. You may find that after losing weight – even before you reach your goal – you may no longer have to depend on diabetes medications.

Better Sleep

Losing weight can also help you to get a better night's sleep. Many people who are very overweight have trouble getting comfortable at night to fall asleep. People also experience problems such as sleep apnea that can be related to weight.

When you follow the hCG diet you can lose weight and get a good night's sleep. You'll be surprised how much better you feel when you're getting the rest you need and deserve. It's hard to believe how much losing weight can improve your sleep.

More Mobility

Many people who are overweight have a hard time exercising. It could be because of pain or because exercising puts a strain on the heart. But when you begin to lose weight – even a little – you'll be able to move much better.

Mobility helps you to perform your daily activities better and to feel like you can take on the world. It will be easier to go grocery shopping, take your kids to the park, and keep your house clean.

It's easy to take those things for granted, but as you feel your mobility improve with your weight loss, you'll begin to see what you were missing before. The hCG diet can help you to get your mobility back and to keep it.

Taking the First Step Toward a Better Life

When you follow the hCG diet, you'll find that you're able to have the life you've always wanted – beyond just the way you look. The hCG diet can help you:

- Lose weight
- Gain self confidence
- Maintain a healthy weight long term
- Create and reach goals
- Reduce your risk of illness
- Reduce your aches and pains
- Feel better about yourself once and for all

Proven Science for a Better Future

The hCG diet is proven scientifically. Since it was developed 50 years ago, millions of people just like you have lost weight and kept it off simply and safely. This method uses hormones that are naturally occurring in your body to boost your metabolism and help your body burn fat.

You won't lose muscle mass, healthy fat, or feel deprived. While you're only eating 500 calories, you won't physically feel hunger. Any hunger you have is probably psychological and a result of the habit of overeating.

Once you've lost weight with the hCG diet, you will understand how effective, simple, and long-lasting it can be. When you follow the program precisely, you'll get the results you've always wanted, but haven't been able to achieve before.

Whether you have five pounds or 105 pounds to lose, you'll find that the hCG diet will work for you. When you follow the simple steps of adding hCG to your body, lowering your caloric intake for 3-6 weeks, and slowly adding sugars and starches back to your diet you'll achieve incredible results.

So, what are you waiting for? It's time to change your life! Begin with talking to your doctor to make sure you're healthy and able to start a diet program. Then start stocking up on supplies and preparing to shed the excess pounds.

You don't want to look back on your life and wonder why you never took the time to lose the weight that's causing you to feel fatigued, in pain, and put you at risk for disease. You want to look back and remember every vital moment you spend feeling full of energy and living your dreams.

Index

Disclaimer

This information is provided as is. The author, publishers and marketers of this information disclaim any loss or liability, either directly or indirectly as a consequence of applying the information and diet information presented herein, or in regard to the use and application of said information. No guarantee is given, either expressed or implied, in regard to the merchantability, accuracy, or acceptability of the information. No information included in this book is meant to be taken as medical advice. It is recommended that you consult your doctor before starting any diet plan. Terms of Use: No information contained in this book should be considered as physical, health related, financial, tax, or legal advice. Your reliance upon information and content obtained by you at or through this publication is solely at your own risk. The author assumes no liability or responsibility for damage or injury to you, other persons, or property arising from any use of any product, information, idea, or instruction contained in the content provided to you through this book.